INDIGENOUS FLOWERS

OF THE

HAWAIIAN ISLANDS

INDIGENOUS FLOWERS

OF THE

HAWAIIAN ISLANDS

FORTY-FOUR PLATES

PAINTED IN WATER-COLOURS

AND DESCRIBED

BY

MRS. FRANCIS SINCLAIR, JR.

"FAIR IS THE BIRD ON LUSTROUS WINGS,
AND STARS THROUGH ALL NIGHT'S SILENT HOURS;
BUT FIRST—OF ALL CREATED THINGS—
IN WONDROUS BEAUTY, STAND THE FLOWERS."

LONDON:
SAMPSON LOW, MARSTON, SEARLE, AND RIVINGTON

1885

Lithographed by

Leighton Brothers, Chromatic Printers,

Stanhope House, Drury Lane,

London.

TO THE HAWAIIAN CHIEFS AND PEOPLE,

WHO HAVE BEEN MY MOST APPRECIATIVE FRIENDS,

AND MOST LENIENT CRITICS.

THIS WORK IS AFFECTIONATELY INSCRIBED.

Privately reprinted by
KvH Publishing
PO Box 881000
Pukalani, Hawai'i 96788

ISBN: 978-1-5323-8758-6

Printed in China.

ABOUT ISABELLA SINCLAIR

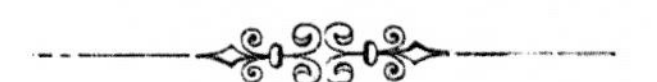

ISABELLA MCHUTCHESON SINCLAIR was born in Scotland in 1840 and migrated with her family to New Zealand in 1861. Her father, William McHutcheson, was the brother of our ancestor Elizabeth ("Eliza") McHutcheson Sinclair. It is likely he moved his family to New Zealand to be closer to his sister, who had migrated from Scotland to New Zealand in 1839 with her husband, Francis Sinclair, and eventually immigrated to Hawaiʻi in 1863.

Francis Sinclair, and his oldest son, George, were lost at sea in 1846 while sailing to Wellington from their home in Pigeon Bay. Adjusting to their tragic loss, Eliza and the five other Sinclair children remained in New Zealand for nearly two decades, ranching and farming until Eliza, at age 63, decided to set sail for new opportunities across the Pacific, settling in the kingdom of Hawaiʻi where the family purchased the island of Niʻihau. Among those traveling with Eliza aboard the Bessie were her daughter Jane and son-in-law Thomas Gay; daughter Helen (Mrs. Charles B. Robinson); and the three unmarried children: James, Francis Jr. (Frank) and Anne, along with five grandchildren.

Two years after arriving in Hawaiʻi, Frank announced his plans to return to New Zealand and marry his cousin Isabella. The couple returned to Hawaiʻi, residing first at Kiekie, Niʻihau, and later at Makaweli, Kauaʻi. Although Isabella and Frank did not have children, Isabella left a priceless legacy with the publication of *Indigenous Flowers of the Hawaiian Islands* in 1885, the first such book with color illustrations of Hawaiʻi's native flora, and today considered to be an extraordinary accounting of Hawaiʻi's flora at the time, much of which has since vanished due to human impact on the island chain's fragile ecosystem.

The primary areas of Isabella's research were the island of Niʻihau and Kokeʻe, Waimea Valley and Olokele Valley on the island of Kauaʻi. In addition to painting the indigenous flowers, she learned the uses and importance of the plants from Native Hawaiians. She also collected botanical specimens, which were sent to Dr. Joseph D. Hooker, director of the Royal Botanical Gardens at Kew, England. Dr. Hooker provided her with each flower's botanical name. Her work and dedication has created a lasting record of the flowers with her beautifully detailed paintings.

This reproduction of Isabella McHutcheson Sinclair's book was published by several great-great-great grandchildren of Isabella's aunt and mother-in-law, Eliza McHutcheson Sinclair. It is dedicated to the rest of our Sinclair clan and future generations, that they too may be inspired by Isabella's work.

INTRODUCTION.

THE Hawaiian Islands have never been celebrated for the beauty of their flora. Indeed, to a stranger, they almost seem destitute of indigenous flowers. But when one comes to search, on hill and plain, sea-coast and cloud-enveloped mountain, it is astonishing the number and variety that are to be found. The islands do not possess many field flowers—in the usual acceptation of the term—but they are rich in flowering trees, shrubs, and vines.

In describing the plants, the seasons are often mentioned. This may be a little confusing to those who are under the impression that the islands possess altogether a tropical climate, with simply a wet and dry season. On the contrary, they enjoy so nearly a temperate climate, that summer and winter, spring and autumn, are pretty distinctly marked. And although plant-life is not nearly so much affected by the seasons, as in colder climates, yet the sleep of winter, and the awakening of spring, are quite apparent.

The following collection of flowers was made upon the islands of Kauai and Niihau, the most northern of the Hawaiian Archipelago. It is not by any means a large collection, considering that the flowering plants of the islands are said by naturalists to exceed four hundred varieties. But this enumeration was made some years ago, and it is probable that many plants have become extinct since then.

The Hawaiian flora seems (like the native human inhabitant) to grow in an easy careless way, which, though pleasingly artistic, and well adapted to what may be termed the natural state of the islands, will not long survive the invasion of foreign plants, and changed conditions. Forest fires, animals, and agriculture, have so changed the islands, within the last fifty or sixty years, that one can now travel for miles, in some districts, without finding a single indigenous plant; the ground being wholly taken possession of by weeds, shrubs, and grasses, imported from various countries. It is remarkable that plants from both tropical and temperate regions seem to thrive equally well on these islands, many of them spreading as if by magic, and rapidly exterminating much of the native flora.

It is only possible to give an approximate definition of the habitat of the various plants; as it often happens, that a lowland plant is found on the mountains, or a mountain plant is found on the coast; but, as a general rule, the habitat given will be found fairly correct. The climate of the islands is greatly modified by the trade-winds. Consequently, plants which are natives of the warm lowlands, may sometimes find a congenial abode at a high elevation, in localities that are sheltered from these cool bracing winds. On the other hand, plants, which are only found on the high cool mountains on the western, warm side of the islands, are often found flourishing near the coast on the eastern, cool side.

In giving the Hawaiian names, great care has been taken to be strictly accurate; but it is often difficult to learn from the ordinary native of the present day, the names of even comparatively common plants, and doubly so with regard to the rarer mountain species. This is partly owing to the changing vegetation, but much more so to the changed habits of the people themselves. Formerly they took a much more intelligent interest in natural objects than they do now, and were nearly all well acquainted with the flora of the islands, and the properties of the various plants, while their former modes of life afforded them many more opportunities and inducements to pursue the study. In olden times, and even to within the last fifty or sixty years, great numbers of the inhabitants went into the mountain districts annually, for various purposes, such as canoe-making, bird-catching, wood-cutting, gathering medicinal herbs, and many other pursuits of pleasure or profit. But of late years the old, healthful, industrious life has so changed, that it is now very seldom they make such expeditions. Consequently it is only from old people—and few even of them—that any reliable information can be obtained, regarding plants which have their habitat far from the abodes of the people of the present generation.

Another difficulty with Hawaiian names,—a difficulty common to all nations without a written language (it was only about fifty years ago that the language was reduced to writing)—is that many plants are known by different names in different districts. But they can generally be traced throughout the islands by a similarity of story and folk-lore connected with each; and many and graceful are these myths, handed down from generation to generation, seemingly from very ancient times. The Hawaiians, like all their race throughout the Pacific, have a most wonderful and intense love of the beauties of nature. Seldom is there found any one more appreciative of the beauty and subtile charm of flowers, than the simple Hawaiian of a generation that is fast passing away.

None can know, save those who have painted flowers in the tropics, the difficulties encountered in obtaining a satisfactory representation, owing to the fragile nature of most of the flowers, and also to the heat of the climate. Many of the flowers fade immediately after they are gathered, and all decay very much sooner than those of temperate climates; thus rendering the utmost expedition necessary—a state of things by no means conducive to the successful execution of the paintings. However, notwithstanding the difficulties encountered, if the perusal of the present work should give as much pleasure as its production has given; or, if it lead others to take up the same wide and delightful study, the author's object will be fully accomplished.

ISABELLA SINCLAIR.

MAKAWELI, KAUAI.

May, 1884.

The Author is indebted to the courtesy and kindness of SIR JOSEPH D. HOOKER for the botanical names of the plants.

I. S.

LONDON, *February*, 1885.

CONTENTS.

CONTENTS.

THE HAU.

Hibiscus tiliaceus, Linn.

THE Hau is found more or less in all parts of the islands from the sea-coast to an elevation of about one thousand feet. When growing singly it attains a considerable size; but where found in groves, it inclines to spread in a dense thicket not over twenty feet high. When the hau grows as a tree, it attains a height of thirty or forty feet, with a short crooked trunk two or three feet in diameter at the base. The tree is a mass of branches and foliage, which renders it an attractive object at all seasons, but especially so in spring and summer, when brilliant with its large handsome, yellow flowers.

The flowers only last one day, opening at sunrise and closing at sunset, and there are no other trees in the islands, and probably few in the world, which produce such a vast number of blossoms in a single season.

The hau is easily grown by simply planting a branch; and, although it matures seed, it is usually propagated by cuttings.

The wood is useful for ox-yokes and other purposes, being light and tough. The inner bark was formerly much used by the natives for making ropes, net-bags, kapa (native cloth), and various other articles. The tree is not peculiar to these islands, being found in most of the Pacific Islands within the tropics. In Tahiti its native name is "purau."

HAU.

LEIGHTON BROTHERS, PRINTERS

PLATE 1.

THE OHIA-LEHUA.

Metrosideros polymorpha, Gaud. var.

Perhaps, there is more graceful song and story connected with the Ohia-lehua than with any other Hawaiian flower. It is so well known, and so striking a feature in the landscape when in full bloom, that it is naturally a favourite with all classes; and is always spoken of with that sort of tenderness with which a Scotchman speaks of heather. If any particular flower were to be taken as an emblem of the Hawaiian Islands; the Ohia-lehua would be most universally chosen, and none would more deserve the honour. It grows to all sizes, from the handsome shrub of fifteen feet high, to the forest tree of a hundred.

On the high table-land of Kauai, about four thousand feet above the sea, a great deal of the lehua is merely a low shrub a few feet in height, but by no means less beautiful than in the lower country. Indeed, the flowers seem to increase in brilliancy in proportion as they are found above the sea level.

The wood is very durable, and useful for various purposes, but, unfortunately, it is not nearly so abundant now as formerly. There are several varieties of the Ohia-lehua, but the present representation is the most common. A few are found almost white or pale yellow. In Spring many of the young leaves are of a bright orange scarlet. These leaves, and afterwards the blossoms, contrast beautifully with the dark evergreen foliage.

The blossoms are loaded with honey, which is the favourite food of the lovely "olokele," a small bird with brilliant scarlet plumage. Few sights in the Hawaiian Islands realise one's dreams of the tropics so fully, as a lehua tree in full bloom, with olokeles flitting from flower to flower, the birds only distinguishable from the blossoms by their quick graceful movements.

In New Zealand there are two varieties of the Ohia-lehua, the "Rata" and "Pohutukawa" bearing the same splendid blossoms as the Hawaiian variety. The rata grows to an immense size, much larger than the lehua. The pohutukawa is comparatively a small tree, rarely exceeding forty feet in height. It is very interesting to find varieties almost precisely similar under such different conditions of climate and in latitudes so far apart.

OHIA-LEHUA.

LEIGHTON BROTHERS, PRINTERS.

THE IEIE.

Freycinetia arborea, Gaud.

THE Ieie has a luxuriance of foliage, and for several months of the year, a brilliancy of colour, that is excelled by few Hawaiian plants. It is a strong climber; its growth seeming to be only limited by the height of the tree upon which it may chance to grow. The main stem, which is about eight inches in circumference, clings to its support by means of ærial roots, and throws out branches every two or three feet. These branches are crowned with graceful plume-like bunches of lance-shaped leaves, from one to two inches broad, and about three feet long. In the centre of each bunch is produced a large, gaudy scarlet inflorescence six or seven inches in diameter, which continues to expand, as the centre, or fruit, increases in size, until it fades, and the fruit—oblong in shape, and from eight to ten inches long—is left to mature. Altogether the ieie is a striking and tropical looking plant; never failing to attract the admiration of all who see it for the first time. It grows on the lowlands in wooded districts, but is found in greatest abundance at an elevation of from one to two thousand feet above the sea. The thick soft bracts of the inflorescence are not unpleasant to the taste, which the rats,—with their usual sagacity—have discovered, and in some localities it is difficult to find a perfect one.

A species of the ieie is found in New Zealand. There the inflorescence is considered, by the natives quite a delicacy. Its Maori name is Kiekie—one of many proofs, of the close connection of the Maori and Hawaiian races.

IEIE.

LEIGHTON BROTHERS, PRINTERS

PLATE 3.

THE PIOI.

Smilax sandwicensis, Kth.

THE Pioi is an upland-vine, plentiful in the forest, a thousand feet and upwards above the sea. It grows strongly, often climbing to a height of thirty or forty feet; throwing out tendrils which clasp anything they touch; and so the plant, spreading from tree to tree, forms very graceful festoons. It blossoms during February and March. The leaves are light glossy green, and the clusters of pale yellow-green flowers have a faint delicate odour. The flowers are followed by bunches of light yellow berries, rather tasteless, but quite agreeable, which are eaten by the natives.

In an ancient myth it is related that the queen and people of a certain island called Ulukaa, used no other food but the pioi—supposing that taro, potatoes, bananas, etc., were poisonous. Through many strange adventures, as related in the myth, a Hawaiian prince came to Ulukaa, and being hungry, gladly ate of his accustomed food, taro, etc., and so taught the queen and her people to use nature's more bountiful gifts.

Pioi.

Leighton Brothers, Printers.

PLATE 4.

THE NUKUIWI.

Strongylodon lucidum, Seem.

THIS is one of the most beautiful of Hawaiian flowers, both in form and colour. The plant is a strong climber, the lower part of the vine being four to six inches in circumference, and so strong that in the old days (when Hawaiians had many games which have now gone out of fashion) it was used for lele-koali, "a swing." When it grows in the woods it does not usually blossom until reaching a considerable height, and its graceful clusters of flowers are generally seen hanging from trees,—upon which the vine has grown,—twenty to thirty feet from the ground.

The flowers are of that peculiarly delicate half-transparent nature, which artists in vain endeavour to represent.

The Nukuiwi grows mostly in warm valleys, from a few hundred to two thousand feet above the sea.

NUKUIWI.

LEIGHTON BROTHERS, PRINTERS

PLATE 5.

THE PUAKAUHI.

Canavalia ensiformis, D. C.?

THE Puakauhi, or awitiwiti, as some of the old natives call it, is a beautiful creeping—or if it chance to find a support—climbing plant. It grows freely on the lowlands in favourable situations, and may occasionally be found on the mountains, but when growing at a greater elevation than a thousand feet, the flowers are far inferior in colour and brilliancy to those found on the lowlands.

It generally blossoms in April and May. The flowers are of a rich dark purple—a colour rare in Hawaiian flowers. They grow in pendent clusters, but not more than three fully open flowers appear at the same time.

Formerly, when the puakauhi was much more plentiful than it is now, the natives prized it for making necklaces and wreaths—mixing it with the blossom of the wiliwili which is of a bright orange scarlet—thus forming a pleasing contrast, very gratifying to their passion for bright colours.

PUAKAUHI.

LEIGHTON BROTHERS, PRINTERS.

THE KOU.

Cordia subcordata, Lam.

THE Kou is strictly a lowland tree, seeming indeed to flourish best close to the sea-coast. It does not attain any great size—seldom being over thirty feet in height, with a short trunk, rarely exceeding two feet in diameter at the base.

The kou is now rare, but it was once a favourite and frequent shade tree, for not only was it quick and easy of growth, but it was the tree from which Hawaiians made most of their handsome and useful wooden dishes. The wood is soft and easily wrought, yet very durable and of a most beautifully variegated brown colour susceptible of a fine polish. In spring many clusters of flowers, of a bright orange colour appear at the end of the branches, gracefully set amid the light green leaves. The flowers are followed by hard black nuts containing an edible kernel.

Fifty or sixty years ago, the kou was comparatively plentiful, but even then it never grew as a forest tree, and was generally found near human habitations. This has led some to suppose that it was imported at a remote period, and the supposition is somewhat strengthened by the fact that the same tree is found on one or two islands south of the equator. This belief, however, is founded upon such slender data, that it is much more probable the kou is truly indigenous; but, from some cause or other, has been gradually decreasing, until, at the present day, it is almost extinct.

Kou.

THE KOKIO-KEOKEO.

Hibiscus Arnottianus, A. Gray, forma.

THIS shrub, or tree, generally grows on the sides of rocky ravines, and is usually found from one thousand to two thousand feet above sea level. It attains a height of about twenty feet, and, when in full flower, is a most beautiful and attractive shrub—the delicate white of the petals, and the pink of the showy stamens, forming a charming contrast with the dark green leaves.

It is quite erratic in its seasons, sometimes blooming late in autumn, and sometimes in spring; being accelerated or retarded by the wetness or dryness of the season. It is to be feared the Kokio-keokeo is doomed to early extinction, as in many places where it was plentiful a few years ago, not a single plant is now to be seen—owing partly to the ravages of cattle and goats, and partly to the changing flora of the islands.

No doubt, at one time it formed a frequent and beautiful feature in the landscape, as it is often mentioned in ancient Hawaiian songs and legends.

KOKIO - KEOKEO

LEIGHTON BROTHERS, PRINTERS.

PLATE 8.

THE KOKIO-ULA.

Hibiscus Arnottianus, A. Gray, forma.

THE remarks which have been made regarding the kokio-keokeo apply equally well to the Kokio-ula, as they are almost identically the same, the only difference being the colour of the flower, which in the case of the kokio-ula is of a brick-red, while the other is pure white.

This variety is perhaps rarer than the white. Both are now very subject to blight. This, and the ravages of cattle and goats already mentioned, will soon make these fine shrubs things of the past.

KOKIO-ULA.

LEIGHTON BROTHERS, PRINTERS

PLATE 9.

THE MILO.

Thespesia populnea, Corr.

The Milo, like the kou, cannot be called a forest tree, as it is seldom found far from the abodes of men, or places which have been inhabited in former times, seeming, like the kou, to have been planted and cared for by human agency. But there is no record or even tradition, of its introduction.

It attains a height of from thirty to forty feet, and is a handsome shade-tree. The leaves are beautifully glossy, and the wind moves them in a most graceful way, somewhat like the quivering of the aspen.

The milo blooms freely, and flowers may be found almost all the year round. They are pale yellow; each petal at its base inside, is purple, but this is rarely noticed, as the flowers do not usually expand more than is shown in the plate.

The tree grows readily from seed upon the lowlands, and seems to resist blight better than most native trees. The wood takes a fine polish, and was used, in former times, by the natives for making calabashes.

MILO.

LEIGHTON BROTHERS, PRINTERS.

THE HAUHELE.

Hibiscus Youngianus, Gaud.

THE subject of the present plate was once a common flower in nearly all valleys, and sheltered places; seeming to flourish equally well on both the leeward and windward sides of the islands. Now cattle and cultivation have almost exterminated the plant on the dry lee-side, but it is still frequently met with on the windward side; where, owing to the more luxuriant vegetation, many plants, which have disappeared from the leeward side, are still found.

The Hauhele was once so plentiful in many parts that the *aho* (thatching sticks) of the houses were made of the stems, and any one who knows what a great quantity of *aho*, a single, old-fashioned house required, will readily see how abundant the plant must have been. It attains a height of from eight to twelve feet, throwing out many branches, which bear showy flowers, and large oblong seed pods. The flowers last but one day. When they open in the morning, they are a very beautiful pink, gradually deepening in colour towards night, when they close. The plant may be found in blossom throughout the year, according to locality, but it mostly blooms in spring and early summer.

The hauhele is armed with minute prickles, which are disagreeable at all times, but particularly so as the season advances, and the seed ripens, for then the prickles are easily detached, and adhere to the hand in a most unpleasant manner.

HAUHELE.

LEIGHTON BROTHERS, PRINTERS

THE KOALI-AWAHIA.

Ipomœa (Pharbitis) insularis, Choisy.

THE Koali-awahia is the most common, and certainly the most beautiful of the convolvulus family in these islands. It is not found in the forest, but almost everywhere else, from the sea-coast to about two thousand feet elevation. It gives, to plain, hill and valley, a bright cheerful aspect by its most charming colour. Unfortunately, cattle and horses destroy it, and the koali-awahia is now confined to places that are more or less protected. Still, the plant is so hardy that it soon reappears wherever the ground is left to nature.

When the flowers open in the morning, they are of a delicate blue purple colour, gradually changing through all the shades of purple, to pink, and finally closing in the evening.

Nothing has a finer effect than grass-land in its original natural state, the tall grass and shrubs festooned with this beautiful blue convolvulus, sparkling with dew in the early sunlight.

The stem and roots of the koali-awahia are much used as a medicine by the natives, and even by many of the foreign residents. Its merits are to allay pain and prevent inflamation. After being pounded into a soft mass, it is applied to all kinds of bruises or broken bones, with wonderfully good effect.

KOALI-AWAHIA.

LEIGHTON BROTHERS, PRINTERS.

PLATE 12.

THE KOALI-AI.

Ipomœa palmata, Forsk.

THE remarks, which have been made regarding the Koali-awahia, apply equally well to the Koali-ai. The former is perhaps the most common, but they are almost always found growing together; and, with their delicate colours and graceful mode of growth, lend a peculiar charm and beauty to all parts of the lowlands that are protected from animals.

The koali-ai, as its name implies (*koali*—the convolvulus plant, *ai*—food), was formerly used by the natives as food, in seasons when their crops of potatoes, etc., failed. The roots and main stems were the parts used, and the taste, though slightly bitter, is not at all unpleasant. The vine is very strong and durable, and is used by the natives for various purposes in place of cordage.

Formerly it was in much request for house-building—every part of a well built native house being fastened with the koali-ai—and, as it was protected from the weather by the thatch, it lasted for a great many years.

The leaves of this convolvulus vary a good deal both in size and form, some are much more palmated than others, but those represented will be found fairly typical.

KOALI-AI.

LEIGHTON BROTHERS, PRINTERS

PLATE 13.

THE PILIKAI.

Ipomœa Turpethum, R. Br. ?

THIS is another of the many varieties of the convolvulus family, which grows so freely all over these islands. As its name indicates (meaning near the sea), it is usually found on the sea-coast; but is also met with in the valleys a considerable distance inland. It grows most abundantly on the windward, or wet side, of the islands. In favourable localities the blossoms and leaves are occasionally larger, but those represented are of average size.

The seeds are held in much repute by the natives as a medicine.

PILIKAI.

THE UALA.

Ipomœa Batatas, Lam.

The Uala, (commonly known as sweet potato), is one of the most widely distributed plants in the world, being found in almost every part of the globe which is at all suitable to its growth. Its natural habitat is a warm climate and rich soil, where it yields immense returns with very little labour. It is either indigenous to the Hawaiian Islands, or was introduced at such a remote period, that all record of the event is lost. In New Zealand, on the contrary (where it is called "Kumara"), there is a distinct tradition of its introduction.

The uala is propagated by planting pieces of the vine, which grow readily, maturing in five or six months, and even in less time, under favourable conditions.

Old natives enumerate nearly fifty varieties, but half that number could not be found in the islands at the present day, as for many years past, the natives have only cultivated the most easily grown kinds.

In Tahiti, the uala—or "umara" as it is there called,—was never prized very highly; but in these islands it has always been considered a superior article of food. On the island of Niihau, where kalo (*Arum esculentum*) is not grown, the uala has always been the principal means of subsistence; and considering that this island was once densely populated, and that the inhabitants are a strong athletic people, we must conclude that the uala is equal, if not superior, to the far-famed kalo, as an article of diet. A field of uala in vigorous growth, entirely covering the ground with dark green leaves interspersed with numerous blossoms, is quite a pretty sight, but unfortunately it is one that is becoming exceedingly rare, as the natives year by year diminish in number, and practice less and less their old healthful modes of life.

UALA.

LEIGHTON BROTHERS, PRINTERS.

PLATE 13

THE POHUEHUE.

Ipomœa pescapræ, Sw.

The Pohuehue is a vigorous and handsome convolvulus, rarely found far from the sea, and generally growing most luxuriantly on the bare sand-hills, immediately above high water mark, where the breakers actually reach its long runners. These runners are often one hundred yards in length, and one root will sometimes cover an acre of ground. Like all the family, it blossoms most profusely in the morning, when it vividly recalls—

> "The lustre of the long convolvuluses
> That coil'd around the stately stems, and ran
> Ev'n to the limit of the land."

The natives use it when fishing, twisting it into long coils, for the purpose of driving the fish into the nets. The seeds are in much repute as a medicine.

The roots and stems are used as food in seasons of great scarcity, but they cause vertigo if eaten exclusively for any length of time, an effect which the leaves also produce upon animals.

The pohuehue is found in most of the South Pacific Islands. In the Society Islands the name is the same as the Hawaiian—a similarity of nomenclature which frequently occurs, and is a strong proof how near the Hawaiians are related to their southern neighbours.

POHUEHUE.

LEIGHTON BROTHERS, PRINTERS

PLATE 16

THE PUAKALA.

Argemone mexicana, Linn. var.

THE subject of the present plate, like all the poppy family wherever found, is a conspicuous and beautiful object in the landscape.

The flower is a perfectly pure white, generally appearing most profusely in February and March, when it covers acres of ground with its delicate snowy bloom. It grows indiscriminately on rich or poor soil, from the sea-coast to a height of about one thousand feet. The leaves and stems, (as the name indicates—*pua*-"flower," *kala*-"rough"), are very prickly, making it a disagreeable plant to handle, and even troublesome to walk through when found growing extensively. It usually attains a height of from three to five feet. The flower is a most difficult one to represent, as it droops immediately after being gathered.

The puakala was noticed and mentioned by Captain Cook, and is one of the few native plants which do not seem to decrease, growing apparently as strongly and profusely now as it did a century ago. The roots contain a large percentage of opium, which the natives formerly used as an opiate in cases of toothache, neuralgia, etc.

The seeds have a wonderful vitality, the plant frequently making its appearance upon land that is allowed to lie fallow—or even upon land over which a fire has passed—where it certainly had not been seen for thirty or forty years, and the seed must have been in the ground for at least that length of time, being too heavy to be carried by the wind.

PUAKALA.

LEIGHTON BROTHERS, PRINTERS

THE WILIWILI.

Erythrina monosperma, Gaud.?

The Wiliwili is found in the driest districts, not only sustaining life, but growing luxuriantly where few other trees could exist. It is seldom more than twenty-five feet high, rarely having a trunk ten feet clear of branches. The wood, when dry, is almost as light as cork, and is much used by the natives for out-riggers to their canoes, on account of its buoyancy. This is one of the few Hawaiian trees, which shed their leaves in autumn. It blooms in early spring, and the tree may often be found covered with flowers while the leaf-buds are only forming.

The flowers vary in colour from pale yellow to orange scarlet. There is no perceptible difference in the trees, but the natives say the wood of those with scarlet flowers is slightly harder and more durable than the other.

The wiliwili carries its seed-pods for many months, and may be seen with a profusion of flowers, together with the seed-pods of the previous season. The pretty bright scarlet seeds were in much request in olden times for *leis*, (necklaces), but like many other graceful Hawaiian fashions, a wiliwili *lei* is now rarely seen.

WILIWILI.

LEIGHTON BROTHERS, PRINTERS.

THE POOLANUI.

Coreopsis cosmoides, A. Gray.

THE Poolanui is a spreading bushy plant, five or six feet high, with twining, interlacing branches, one plant covering from eight to ten feet of ground. It generally grows under the shade of open forest, in the mountain regions at various heights above the sea, but seldom less than two thousand feet.

In ordinary seasons, it blooms in April and May, but occasionally flowers may be found as late as the end of June or beginning of July.

The poolanui is quite a striking flower, not only on account of its size and colour, but also on account of the great number in bloom at the same time—giving the sombre forest quite a bright appearance during the spring months. It is a useful fodder plant, cattle and horses eating it with avidity, but it soon disappears if constantly eaten down.

The poolanui seems a flower that might be improved by cultivation, and would doubtless grow from seed under suitable conditions.

POOLANUI.

LEIGHTON BROTHERS, PRINTERS.

PLATE 19.

THE UKIUKI.

Dianella ensifolia, Red.

THE subject of the present plate is a member of the large and varied family of lilies. Its pretty berries, and graceful leaves marked with bright bits of colour, render the Uki (as the natives generally call it) a prize to the artist, and a favourite with all who take an interest in the beauties of nature. The flowers are small and do not attract notice. It is the berries,—which become a bluish purple when ripe,—that give the plant its chief attraction, contrasting charmingly with the yellow, unripe berries and the bright green leaves.

The uki grows on the high lands, the cool air of the mountains seeming a necessity of its existence. Yet it is not often found in the very damp sunless regions of the interior; its habitat being the slopes of mountains, where the ground is dry and rain not very frequent. It usually grows in isolated bunches, about three feet high, and, like the majority of Hawaiian plants, has feeble roots, and is easily destroyed.

The berry is a favourite food of the wild fowl; but the natives do not eat it, as they say it contains certain properties poisonous to the human system.

When the natives are upon mountain expeditions, they often use the uki for thatching their temporary huts, a purpose for which it is well adapted.

UKIUKI.

LEIGHTON BROTHERS, PRINTERS.

THE NEHE.

Lipochœta australis, A. Gray, var.

THE Nehe was a very common plant a few years ago. Now, it is rarely found on any spot that is accessible to cattle. It is a spreading, bushy plant, three or four feet high, bearing quantities of yellow flowers all through spring and summer.

Although it seems equally at home on low or high land, yet, when growing upon the mountains, the flowers are much larger, brighter coloured, and finer in every way, than those found near the coast. The present representation was painted about three thousand feet above the sea, and, to those who have only seen the plant near the coast, might seem another species, but in reality it is the same.

The leaves of the nehe are sometimes used like the common tea, by both foreigners and natives, and have rather a pleasant flavour, besides possessing, it is said, some medicinal value.

NEHE.

LEIGHTON BROTHERS, PRINTERS

THE OHAI.

Sesbania (Agati) tomentosa, A. Gray.

The Ohai is a beautiful shrub, bearing graceful clusters of dark orange-scarlet flowers. It is a native of the lowlands on the leeward sides of the islands, where it flourishes best upon ground that is partially flooded by the heavy rains of winter. It is a rapid growing plant, attaining a height of eight to twelve feet in a single season. It grows freely from seed, making its appearance immediately after the first heavy autumn rains, and by the second month of spring it is in full bloom, and continues to blossom each spring for a few years; but the flowers are finest during the first season.

The seeds are contained in long slender pods, which may be found hanging on the same branch with the bright clusters of blossoms, and rather adding than detracting from their beauty.

The ohai was once plentiful, growing singly and in groves; but since the introduction of cattle the plant has almost disappeared.

OHAI.

LEIGHTON BROTHERS PRINTERS.

PLATE 22.

THE MAO.

Gossypium tomentosum, Nutt.

THE Mao is a species of the plant which produces the well known cotton of commerce. It is a hardy bushy shrub, five to eight feet high, only found on the lowlands. It is not so plentiful now as formerly, and is hardly ever met with excepting in protected places. The leaves are of a bluish green, a tint which is rather uncommon. The flowers are of the beautiful colour known as primrose, which, as all who have painted flowers are aware, is a most difficult colour to represent at all resembling nature.

The Mao blossoms, more or less, all the year round; therefore, both seed and flowers are found simultaneously.

Altogether, it is an attractive shrub; the leaves, flowers, and seed, forming a variety of colour, which is very pleasing. It will be observed that the bolls represented are of a brown colour, instead of the snowy white of the sea-island, and other varieties of the cotton of cultivation.

MAO.

THE AEAE.

Lycium sandwicense, A. Gray.

The Aeae is found upon low-lying damp ground on the margin of salt lagoons. It is quite independent of rain, as it only grows where its roots are nourished by the brackish water filtering through the soil. In the dry districts it is a valuable plant for live stock, its thick juicy leaves affording a green wholesome "bite," when nearly all other plants are burnt up with the dry hot weather of summer and autumn.

If standing singly, the aeae grows as a pretty little upright bush, two or three feet high, but when found in great quantities close together, it inclines to grow prostrate, and the branches become interlaced, forming thickets, which are almost impenetrable.

The flowers are small, of a pale lilac colour, and not very noticeable, but the bright scarlet berry—about the size of a currant—is very pretty, and is one of the few edible berries in the islands. It is quite pleasant and refreshing to the taste, with a peculiar saline flavour, by no means disagreeable.

The aeae has no stated time of blooming. Blossoms and fruit may be found, more or less, at all seasons.

LEIGHTON BROTHERS, PRINTERS

AEAE.

PLATE 24.

THE KAUILA.

Alphitonia excelsa, Reiss.

THE Kauila is a stately forest tree, from seventy to eighty feet in height. The wood is a deep red colour, very fine in the grain, and is perhaps the most beautiful of all Hawaiian woods.

It was never a plentiful tree. Now it is extremely rare. It is mostly found on the lee-sides of the islands, from two thousand to three thousand feet above the sea. The leaves are dark glossy green, the under-side much lighter than the upper, and when the wind moves them, the blending of the different shades of green is very pleasing. The blossoms are small and by no means showy when studied singly, but when the tree is in full blossom the effect is very fine.

The kauila was highly prized by the natives in olden times, on account of the beauty and durability of the wood. It was much used for spears, mallets, etc. It was also valuable for house building.

The Hawaiian mode of building, was to set the *pou* (side-posts) in the ground; therefore a durable wood was of great importance. When a native once succeeded in building a house of kauila he felt secure for his lifetime; the thatch being the only part that required renewing.

KAUILA.

LEIGHTON BROTHERS, PRINTERS.

THE KOLOKOLO.

Vitex trifolia, Linn. var. *unifoliolata.*

THE Kolokolo is a pretty, and at the same time, a useful plant; as it helps to bind the loose sand-hills, with its numerous spreading stems and roots. The main stems are creeping, and at short intervals send up leafy stalks about two feet high, the flowers and seeds being produced at the top as shown in the plate. The kolokolo is sometimes found growing upon soil, but pure sand is its favourite locality. It is a hardy plant, suffering little from drought, and blooming more or less all the year round; but most profusely in spring, when it is a very attractive object, covering the white sand with its purple flowers and bright green leaves. The leaves have a pleasant aromatic taste and odour, not unlike sage, for which they are by no means a bad substitute.

KOLOKOLO.

LEIGHTON BROTHERS, PRINTERS.

PLATE 26.

THE KOLOKOLO-KUAHIWI.

Lysimachia Hillebrandi, Hook. fil.

THIS plant is only found on the high lands of the interior, from three to five thousand feet above the level of the sea. Its graceful, drooping flowers grow on woody stems one to three feet in height.

Often hidden by tall grass or rushes, the flower might readily be passed by without notice, but, when once known, few Hawaiian flowers are more prized; not only for its beauty of shape and colour, but also for its exquisitely delicate odour.

This flower is familiar to few of the foreign residents, and even a great proportion of the natives have never seen it, as they seldom now-a-days visit the mountain districts where it grows, thus it is truly one of nature's gems

— "Born to blush unseen
And waste its sweetness on the desert air."

There is an interesting superstition regarding the Kolokolo-kuahiwi, which is as follows. If the flower is pulled, "the tears of heaven" (rain) will fall. So in the old days the natives were careful not to gather it, as they dreaded the cold mountain rain, which was very inconvenient during their expeditions.

LEIGHTON BROTHERS, PRINTERS.

PLATE 27.

KOLOKOLO - KUAHIWI.

THE NANEA.

Vigna lutea, A. Gray.

THE Nanea is a vine which was once plentiful on the lowlands. Now it is very rare. It is not a deeply rooted plant; therefore it readily gives way to the more vigorous growth of imported weeds and grasses.

When growing in a favourable situation, the leaves, which are of a pleasant refreshing green, quite cover the ground, and they, together with the bright little flowers, form a pleasing picture.

The nanea is a perennial plant, and blooms freely during spring and summer. The seed-pods may be found on the plant in all stages of development, simultaneously with the blossom. The young pods are light green, gradually changing to brown, and becoming almost black, when fully ripe.

THE HUNAKAI.

Ipomœa (Batatas) acetosæfolia, Choisy.

THE Hunakai, as its name (sea-foam) indicates, grows on the actual margin of the ocean, where the wreck and drift-wood are cast by every high surf; and under no circumstances is it ever found far from the influence of the salt sea-air. The vine grows rather peculiarly, running an inch or two under the sand, and at short intervals throwing up flowers and little clusters of leaves. The flowers, although beautiful, are not nearly so attractive when studied singly, as when seen covering the sand with their delicate bloom set in small clusters of bright tinted leaves.

Like many maritime plants, the hunakai blossoms more or less all the year round, depending greatly on the wetness or dryness of the season.

LEIGHTON BROTHERS, PRINTERS.

PLATE 28.

THE HOI.

Dioscorea sativa, Linn.

THE Hoi is quite a peculiar vine. A curious potato-like tuber grows at the nodi of the stem, simultaneously with the blossoms. These blossoms are very minute, growing on a gracefully pendulous stem ten to fifteen inches in length, which, together with the large dark green leaves, and peculiar tuber above mentioned, give the plant quite a unique appearance. Formerly, in times of great scarcity of food, the hoi was cooked and eaten by the natives. It has a very bitter taste, and was only used in extreme cases of famine. Some of the vines are much larger than the piece represented, the leaves being often twelve inches across, and the tuber three inches in diameter. Its favourite locality is rough rocky ground, and it has the peculiarity of growing equally well on the lowlands and at an elevation of a thousand feet.

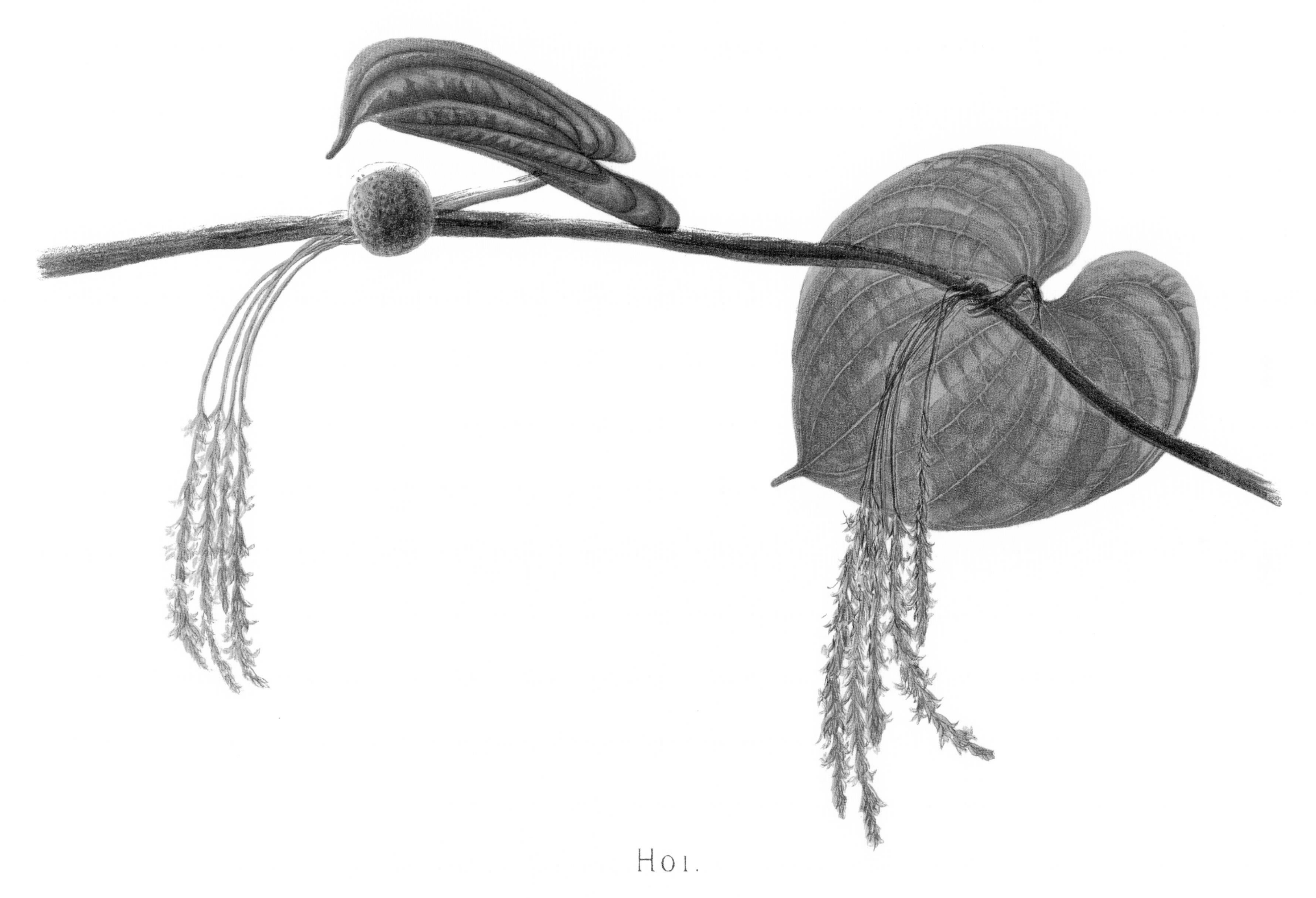

HOI.

LEIGHTON BROTHERS, PRINTERS

PLATE 29.

THE NOHU.

Tribulus cistoides, Linn.

THE subject of the present plate usually grows near the sea; and although it may occasionally be found a few miles inland, yet it is only upon the sea coast that it attains full beauty of form and colour. The flowers are in greatest perfection soon after sunrise, when they present a beautiful field of golden-yellow, filling the air with a delicate fragrance. They close at sunset and open at sunrise, generally lasting only two days; but there is no diminution of bloom, as there is a daily succession of flowers as long as the weather continues favourable; for like many of the lowland plants, the Nohu is more influenced by the weather than by the seasons.

The nohu is a prostrate plant, and where it grows vigorously, quite covers the ground with a close green carpet, very pretty to look at but by no means pleasant to walk upon—unless one is well shod—on account of the sharp strong thorns with which the seed-pods are armed; and as the natives generally go bare-foot, the nohu, in spite of its charms of colour and fragrance, is not a favourite with them.

NOHU.

THE KAKALAIOA.

Cœsalpinia Bonducella, Flem.

The name of this plant means "thorny," and it is a most appropriate and descriptive one, as those who have had their clothes torn and hands scratched by its sharp hooklike thorns will at once admit. Yet, like all nature's productions, the Kakalaioa has its redeeming features. With its graceful inflorescence and foliage, it covers many a spot that would otherwise be barren, and is always a pleasing object, if one has the common sense to leave it alone.

It generally grows in rocky places on the lowlands, and by its strong rambling mode of growth covers everything within its reach.

The plant is altogether too large and spreading to be well represented—only part of the leaf, which is bipinnate, is shown.

The seed-pods are very curious, being thickly covered with sharp spines, and are first green, then brown, and when ripe, almost black. They grow in bunches of from eight to thirteen pods—only two are represented.—When fully ripe, these burst open, displaying from two to four, round, very hard seeds, about the size of marbles, which the native boys use in their games instead of the genuine article.

Kakalaioa.

Leighton Brothers, Printers.

THE NAUPAKA.

Scævola Kœnigii, Vahl. var.

THE Naupaka is a low spreading shrub, with thick soft branches and fleshy leaves. It is always found close to the sea, often within reach of the waves. In sheltered situations it attains a height of five or six feet, but generally it is not more than three feet high. It bears the scorching heat of summer in a wonderful manner, and in the driest seasons may be seen with bright green foliage, when most other plants are bare and brown ; thus giving the sandy beaches where it grows, a fresh, pleasant appearance.

The flowers and berries are not by any means conspicuous, but are quite pretty when examined closely.

NAUPAKA

LEIGHTON BROTHERS, PRINTERS

THE OHENAUPAKA.

Scævola glabra, H. & A.

THIS is a native of the far misty mountains, from four to five thousand feet above the level of the sea; where for more than half the year it is wet with the mountain rains, and enveloped in the trade-wind clouds.

The shrub attains a height of ten to fifteen feet, growing in the dense woods, evidently adapted to the humid and cold atmosphere of the bleakest mountain districts. There it may be seen upon the edge of precipices, exposed to the fierce sweep of the trade-winds, which gather strength as they rush upward from the ravines thousands of feet below.

OHENAUPAKA.

LEIGHTON BROTHERS, PRINTERS

THE ILIAHI.

Santalum ellipticum, Gaud.

ILIAHI—the far-famed sandalwood—is widely distributed throughout the islands of the Pacific, and fifty years ago was a valuable article of commerce. Immense quantities were consumed in China, being burnt as incense before sacred shrines and idols, besides being used in the manufacture of various fancy articles. For many years, the iliahi was one of the principal sources of revenue of the Hawaiian kings and chiefs. So vigorously did they prosecute the business of cutting and exporting it, that they exhausted the supply, and to-day it is a very rare tree, although frequently found as a shrub.

The iliahi, in its natural state on the Hawaiian islands, was usually a straight handsome tree, from fifty to eighty feet high, and from two to three feet in diameter at the base. The wood is hard, of a light brown colour, and retains its scent in a wonderful manner, even small pieces being quite fragrant after a lapse of forty or fifty years. The older the tree the more valuable it becomes, as the fragrance increases with age.

The flower is by no means striking, but in a collection of Hawaiian plants the iliahi is too historical to be passed over.

ILIAHI.

LEIGHTON BROTHERS, PRINTERS

PLATE 34

THE NOHUANU.

Geranium cuneatum, Hook. var.

THIS little flower is found on the high, bleak swamp-land of the interior, at an elevation of about four thousand feet above the sea. The plant is a bushy little shrub, a foot or so in height, and usually grows among tall grass and rushes. By its mode of growth and general appearance, the Nohuanu may be considered one of the few field-flowers which the islands possess.

Few natives, and still fewer foreigners, are acquainted with the flower, as its habitat is seldom visited by man. There are so few flowers in the islands which correspond to what are familiarly termed field-flowers, that the nohuanu cannot fail to recall to the wanderer from other lands the tenderly remembered flowers of his childhood, "Gathered when life was new."

NOHUANU.

LEIGHTON BROTHERS, PRINTERS.

THE PUAHANUI.

Broussaisia pellucida, Gaud.

THIS plant is only found upon the wooded mountains from three thousand to four thousand feet above the sea, where it grows under the shade of the dense forest, and where the vegetation is almost constantly kept damp by rain or mist.

It is a pretty, thrifty looking shrub, about seven or eight feet high, and its large clusters of purple blossoms show to much advantage against the dark green leaves.

Those conversant with Hawaiian plants may know the Puahanui under some other name, as it is one of those plants, which, being rather uncommon, are often confused with something else. This confusion could not have occurred in olden times, when the great majority of the people knew the names, and understood the properties of every plant. Now one seldom meets a native who can give any reliable information on the subject.

PUAHANUI.

LEIGHTON BROTHERS, PRINTERS.

PLATE 36.

THE AKAAKAAWA.

Hillebrandia sandwicensis, Oliv.

THIS is one of the beautiful begonia family; and, like many of its species, has that peculiar delicacy—almost transparency—which, as has been remarked before, renders a flower so difficult to delineate. It is a graceful plant from three to four feet high, spreading from the root into many soft green stems with a profusion of conspicuously large leaves, each stem producing several clusters of delicate pink and white blossoms.

It is found in greatest profusion in shady and humid mountain ravines, near the misty spray of waterfalls, where its delicate clusters of flowers and large leaves form a picture of great beauty; especially when seen through the faint rainbows, which are almost constantly floating in the spray of tropical waterfalls.

AKAAKAAWA.

THE HIALOA.

Waltheria americana, Linn.

The Hialoa is rather an insignificant looking plant, and is apt to be passed without notice; but upon examination it will be found well worthy of a closer acquaintance. The hialoa appears to greatest advantage in early spring; for although it bears the hot dry weather better than many plants, yet it assumes in summer a faded and somewhat shabby appearance. But in early spring, with its exquisite bits of colour, and soft corrugated leaves, it is a very pretty plant indeed; though one of Hawaii's humblest.

The hialoa grows everywhere on the lowlands; and varies from two to six feet in height. The plant contains a great quantity of gluten, which the natives turned to account in their primitive days, using the pounded leaves for filling the seams and cracks of their canoes.

HIALOA.

LEIGHTON BROTHERS, PRINTERS

THE AALII.

Dodonœa viscosa, Linn.

THE Aalii is mostly found in the dry districts of the islands, often sustaining life and vigour on the most arid spots. On the lowlands, and within three or four miles of the sea, it is merely a shrub from four to eight feet high, but on the mountains it sometimes attains a height of thirty feet, with a trunk three feet in circumference. Upon the low dry land, it usually grows bushy, and in groves, quite shading the ground with its thick foliage, very grateful to the eye in the blazing sunshine. The blossom is small and insignificant, but the outer covering (represented in the plate) of the seed, is pretty and bright coloured, hanging in graceful clusters, reminding one very much of hops.

The wood of the Aalii is exceedingly hard and susceptible of a fine polish. It is almost identical with the *akeake* of New Zealand, so much so that the most casual observer cannot fail to perceive the similarity. In olden times it was used by the Hawaiians for making spears and others implements of war—a purpose for which the *Maories* used the *akeake*.

AALII.

LEIGHTON BROTHERS, PRINTERS.

PLATE 39.

THE NONI.

Morinda citrifolia, Linn.

THE Noni, although usually a shrub not exceeding ten or twelve feet in height, sometimes attains the dimensions of a tree, and is found—under favourable circumstances—from twenty to thirty feet high. It is wonderfully prolific and hardy, being generally loaded with fruit and leaves at all seasons of the year, irrespective of rain or sunshine.

The fruit and foliage are much more attractive than the flowers, which are small and dull in colour. But it is only in appearance that the fruit is inviting, being acrid to the taste, yet, in times of scarcity, the natives not only eat it in a raw state, but by cooking make it, not exactly nice, but much less disagreeable.

The tree begins branching from the ground, and presents a mass of large bright green leaves intermingled with the fruit in all stages of growth. The inflorescence of the noni is peculiar, as will be noticed in the plate. From three to six of the small flowers appear, lasting for a day or two, and gradually dropping off as the fruit increases in size. Meanwhile, another embryo fruit appears nearer the end of the branch, and so the wonderful process goes on, month after month, year after year, seemingly ad infinitum.

NONI.

LEIGHTON BROTHERS, PRINTERS

PLATE 40.

THE OHIA-AI.

Eugenia (Jambosa) malaccensis, Linn.

THE Ohia-ai is mostly found in sheltered valleys near streams. It is easily propagated by seed, and grows rapidly, beginning to bear fruit when seven or eight years old. Where isolated it inclines to grow like a shrub, but, when growing in groves, or sheltered by other trees, it often attains a height of fifty feet, with a perfectly straight trunk.

It is a handsome tree at all times, but especially so in spring, when brilliant with masses of splendid carmine blossoms; and again in autumn, when loaded with fruit of the same beautiful colour.

The flowers grow directly from the branches, and even from the trunk, giving the tree a peculiarly rich appearance; and the blossoms are so numerous, that, when they fall, the ground beneath is covered with a bright red carpet.

The fruit in shape resembles an oblong apple, with a white, juicy, but rather insipid pulp.

The ohia-ai is also a native of the Society Islands; where it is called *ahia*, and is similar in all respects to the Hawaiian variety.

OHIA-AI.

LEIGHTON BROTHERS, PRINTERS

PLATE 41.

THE PUAPILO.

Capparis sandwichiana, D. C.

THIS is one of the most beautiful, and characteristic of Hawaiian flowers. But one of the most difficult to represent, owing to its fragile nature, and also on account of it being night-blooming. The blossoms open at sunset and wither soon after sunrise. Under these circumstances, it is no easy task to accomplish a painting of the Puapilo. The only way of doing so is to begin work at the earliest break of dawn, and even then the greatest expedition is necessary, as in an hour or so, the flowers not only fade, but lose their delicate creamy white, and become of a dull pink colour.

The plant has no stated time of blooming. Flowers can be found at almost any season of the year, but in greatest profusion soon after the autumn rains. The plant is found on the lowlands upon broken rocky ground, its branches spreading over and around the stones, and it may frequently be seen growing on the face of perpendicular cliffs, with seemingly no soil whatever to nourish its roots. Perhaps this peculiarity of growing on inaccessible places has saved it from the ravages of animals, which have destroyed so many indigenous plants.

PUAPILO.

LEIGHTON BROTHERS, PRINTERS

PLATE 42

THE AKALA.

Rubus hawaiensis, Gray. ?

The subject of the present plate is a large and handsome variety of the common raspberry. It attains a height of from ten to fifteen feet, and produces a profusion of large beautiful purple berries; which become almost black when fully ripe. Unfortunately, the fruit is not so pleasing to the taste as to the eye, being insipid and flavourless, but its appearance is so tempting, that one cannot refrain from gathering the delicious looking berries. The Akala is generally found growing near streams, or on damp ground, far in the cool mountains. It is rarely seen at a less elevation than three thousand feet. Therefore, its habitat may be considered a temperate climate. It blossoms in April and May, and is generally in full bearing about the middle of June, but like many plants of the islands, its flowering and fruiting seasons vary considerably.

AKALA.

LEIGHTON BROTHERS, PRINTERS.

PLATE 43.

THE PAPALA.

Charpentiera ovata, Gaud. var.?

WHEN in full bloom, the Papala has a very peculiar and graceful appearance. The blossom, by its form and colour, reminds one of the most delicate seaweed.

It comes into flower in April and May, generally continuing in blossom throughout the summer. The tree attains a height of about twenty feet, and grows only upon the highlands from two to three thousand feet above the sea.

The wood is very light and porous, and being easily ignited, is used by the natives for most original and grand displays of fireworks. On the north-west side of Kauai the coast is extremely precipitous, the cliffs rising abruptly from the sea to a height of from one to two thousand feet, and from these giddy heights, the ingenious and beautiful pyrotechnic displays take place. On dark moonless nights, upon certain points of these awful precipices—where a stone would drop sheer into the sea—the operator takes his stand with a supply of papala sticks, and lighting one, launches it into space. The buoyancy of the wood, and the action of the wind sweeping up the face of the cliffs, cause the burning wood to float in mid-air, rising or falling according to the force of the wind, sometimes darting far seaward, and again drifting towards the land. Firebrand follows firebrand, until, to the spectators (who enjoy the scene in canoes upon the ocean hundreds of feet below) the heavens appear ablaze with great shooting stars, rising and falling, crossing and recrossing each other, in the most weird manner. So the display continues until the firebrands are consumed, or a lull in the wind permits them to descend slowly and gracefully to the sea.

PAPALA.